RAILWAY INDUSTRY STANDARD

OF THE PEOPLE'S REPUBLIC OF CHINA

Rail Crane Flatcar

TB/T 3194-2008

Issued by Ministry of Railways of the People's Republic of China

Issued Date: March 14, 2008

Valid Date: July 1, 2008

China Railway Publishing House

Beijing 2019

图书在版编目(CIP)数据

起重轨道平车:TB/T 3194—2008:英文/中华人民共和国国家铁路局组织编译.—北京:中国铁道出版社,2019.1
ISBN 978-7-113-25471-1

Ⅰ.①起… Ⅱ.①中… Ⅲ.①铁路起重机-国家标准-中国-英文
Ⅳ.①TH218-65

中国版本图书馆 CIP 数据核字(2019)第 018553 号

Chinese version first published in the People's Republic of China in 2008
English version first published in the People's Republic of China in 2019
by China Railway Publishing House
No. 8,You'anmen West Street, Xicheng District
Beijing, 100054
www. tdpress. com

Printed in China by Beijing Hucais Culture Communication Co.,Ltd.

ISBN 978-7-113-25471-1

Introduction to the English Version

To promote the exchange and cooperation in railway technology between China and the rest of the world, Planning and Standard Research Institute of National Railway Administration, entrusted by National Railway Administration of the People's Republic of China, organized the translation and preparation of Chinese railway technology and product standards.

This standard is the official English version of *Rail Crane Flatcar* (TB/T 3194-2008). The original Chinese version of this standard was issued by the former Ministry of Railways of the People's Republic of China and came into effect on July 1st, 2008. In case of discrepancies between the two versions, the Chinese version shall prevail. National Railway Administration of the People's Republic of China owns the copyright of this English version.

Your comments are invited for next revision of this standard and should be addressed to Technology and Legislation Department of National Railway Administration and Planning and Standard Research Institute of National Railway Administration.

Address: Technology and Legislation Department of National Railway Administration, No. 6, Fuxing Road, Beijing, 100891, P. R. China.

Planning and Standard Research Institute of National Railway Administration, B block, JianGong Building, No. 1, Guanglian Road, Xicheng District, Beijing, 100055, P. R. China.

The translation was performed by Li Mengfei and Luo Jun.

The translation was reviewed by Gan Jiandong, Wang Shuai, Zhou Bian, Zhang Lizhi, Zhang Ying, Liu Gui, Zhang Dianjun, Liang Yiru, Xiao Fei and Zhao Lin.

Notice of National Railway Administration on Issuing the English Version of Eighteen Chinese Railway Technical Standards including *Specification of Alarm Device for Monitoring Locomotive Bearing Temperature*

Document GTKF〔2018〕No. 58

National Railway Administration of the People's Republic of China organized the translation and preparation of eighteen Chinese railway technical standards including *Specification of Alarm Device for Monitoring Locomotive Bearing Temperature* (TB/T 3057-2002). In case of discrepancies between the two versions, the Chinese version shall prevail.

China Railway Publishing House is authorized to publish and distribute the English version of these standards.

Attached here is a list of the English version of these standards.

S/N	Chinese title	English title	Standard number
1	机车轴承温度监测报警装置技术条件	Specification of Alarm Device for Monitoring Locomotive Bearing Temperature	TB/T 3057-2002
2	铁道客车用集中轴温报警器	Centralized Journal Temperature Alarm System for Railway Passenger Car	TB/T 2226-2016
3	标准轨距铁路道岔技术条件	Technical Specifications for Turnouts for Standard-gauge Railway	TB/T 412-2014
4	高锰钢辙叉技术条件	Technical Specifications for Solid Manganese Steel Frogs	TB/T 447-2004
5	铁路道岔用非对称断面钢轨	Asymmetric Cross Section Rails for Railway Turnouts	TB/T 3109-2013
6	43 kg/m～75 kg/m 钢轨接头夹板订货技术条件	Technical Specifications for Procurement of 43 kg/m-75 kg/m as Joint Bar	TB/T 2345-2008
7	防腐木枕	Preservatively-treated Wood Sleepers	TB/T 3172-2007
8	铁路碎石道床底碴	Railway Sub-ballast	TB/T 2897-1998
9	预制先张法预应力混凝土铁路桥简支 T 梁技术条件	Technical Specifications for Precast Pretensioned Prestressed Concrete Simple-supported T-girders for Railway Bridges	TB/T 2484-2005
10	铁路桥梁球型支座	Spherical Bearings for Railway Bridges	TB/T 3320-2013
11	铁路桥梁板式橡胶支座	Elastomeric Bearings for Railway Bridges	TB/T 1893-2006
12	轻型轨道车技术条件	Technical Specifications for Light Duty Railway Trolley	TB/T 2139-2004
13	起重轨道车	Track Crane	TB/T 2187-2014
14	起重轨道平车	Rail Crane Flatcar	TB/T 3194-2008
15	铁路车站计算机联锁技术条件	Technical Specifications for Computer Based Interlocking	TB/T 3027-2015
16	轨道电路通用技术条件	General Technical Specifications for Track Circuit	TB/T 2852-2015

(continued)

S/N	Chinese title	English title	Standard number
17	铁路信号计轴应用系统技术条件	Technical Specifications for Railway Signal Axle Counter Application System	TB/T 3189-2007
18	铁路声屏障声学构件技术要求与测试方法	Technique Requirements and Measurement of Acoustic Elements of Railway Sound Barrier	TB/T 3122-2010

National Railway Administration of the People's Republic of China
July 10, 2018

Contents

Foreword

This standard is proposed and managed by the China Railway Engineering Machinery Research & Design Institution Co. , Ltd.

This standard is drafted by China Railway BAOJI Machinery Co. , Ltd. , Xiang Fan Jinying Railway Vehicle Co. , Ltd. , China CREC Railway Electrification Bureau Group and the China Railway Engineering Machinery Research & Design Institution Co. , Ltd.

This standard is mainly drafted by Zhang Dianjun, Li Ling, Liang Yiru, Chen Bin and Yu Hongwei.

Rail Crane Flatcar

1 Scope

This standard specifies the operation conditions, technical parameters and series, requirements, test methods, test rules, marking, packaging, transportation and storage etc. for rail crane flatcar (Hereinafter referred to crane flatcar).

This standard is applicable to the design, manufacture and test of newly-built standard gauge crane flatcar without own power and with a maximum rated lifting capacity of no more than 16 t. The crane flatcar with rated lifting capacity of 20 t to 50 t, two-axle crane flatcar and non-standard gauge crane flatcar can also refer to related clauses of this standard.

2 Normative References

The following normative documents contain provisions which, through reference in this text, constitute provisions of this standard. For dated references, subsequent amendments (except for the corrigendum), or revisions, of any of these publications do not apply to this standard. However parties to agreements based on this standard are encouraged to investigate the possibility of applying the most recent editions of the normative documents indicated below. For undated references, the latest edition of the normative document referred to applies.

GB 146.1 *Rolling stock gauge for standard gauge railways*

GB/T 3766 *General technical conditions in the hydraulic system*

GB/T 3811 *Design rules for cranes*

GB/T 6067 *Safety rules for lifting appliances*

GB/T 6068.2-2005 *Test code for truck crane and mobile crane*

GB/T 12602 *Lifting appliances-safety devices against overloading*

GB/T 13306 *Plates*

GB/T 14039-1993 *Hydraulic fluid power-Fluids-Method for coding the level of contamination by solid particles*

GB/T 14097-1999 *Noise limits for small and medium power diesel engines*

GB/T 15052-1994 *Cranes-Safety signs and hazard pictorials-General principles*

GB/T 16904.1 *Checking of rolling stock clearance for standard gauge railways-Part 1: Methods for inspecting*

GB/T 17426 *Dynamic performance evaluation and test method for particular class vehicles and tracked machine*

TB/T 1580 *Technical requirements for newly build locomotive and vehicles welding*

TB/T 1854-2006 *Draw up method for rail permanent way machinery*

TB/T 2033 *General technical specification for rail flat car*

TB/T 2187 *General technical specification for railway crane*

TB/T 2879.5-1998 *Rolling stock Painting and coating part 5: Technical specification of protection and coating for passenger coach and traction motive vehicle*

TB/T 2911-1988 *General technical specification of vehicle riveting*

TB/T 3021-2001 *Electric apparatus of rolling stock*

3 Operation Conditions

3.1 Ambient temperature: −25 ℃ to 45 ℃.

3.2 Ambient humidity: relative humidity no more than 95%.

3.3 Altitude: no higher than 1 000 m.

3.4 Max. wind speed: not exceeding 13.8 m/s.

3.5 Min. radius of curve negotiated: not smaller than 145 m.

3.6 Line gradient: not greater than 25‰.

3.7 Track with good maintenance, flat and solid track shoulder and non-sink rail base during operation.

3.8 Special requirements of operation condition shall be agreed between the user and supplier.

4 Technical Parameters and Crane Flatcar Series

4.1 Gauge:1 435 mm.

4.2 Number of axles: 4.

4.3 Wheel diameter: should be ϕ840 mm.

4.4 Max. running speed: not smaller than 100 km/h.

4.5 Height of coupler centerline above the top of rail: 880 mm±10 mm.

4.6 Power supply voltage: DC 12 V or DC 24 V.

4.7 Crane Flatcar Series

Type of the crane flatcar shall be compiled according to the provisions of TB/T 1854-2006. Crane flatcar series shall be designed according to the rule of serialization which is made up of maximum rated lifting capacity and corresponding minimum working outreach. The value is shown in Table 1.

Table 1

Maximum rated lifting capacity t	Maximum rated load working outreach(no less than) m
4	3
5	3
6.3	3
8	3.5
10	3.5
16	3.5

5 Requirements

5.1 General Requirements

5.1.1 The upper crane design shall comply with the requirements of GB/T 3811 and the lower rail flatcar design shall comply with the requirements of TB/T 2033 and TB/T 2187 and the crane flatcar shall be designed according to the rules of standardization, serialization, interchangeability and maintainability.

5.1.2 The crane flatcar shall be manufactured, procured and assembled by product drawings and technical documents which have been approved by the specified procedure.

5.1.3 The general layout of the crane flatcar shall allow each component to be easily disassembled, adjusted and repaired.

5.1.4 The welding, inspection and acceptance of the welding seam of the crane flatcar shall comply with the requirements of TB/T 1580, GB/T 6067 and product drawings.

5.1.5 The riveting structure quality shall comply with the provisions of TB/T 2911-1998.

5.1.6 The oil pipe, air pipe, wire and cable of the crane flatcar shall be laid reasonably, arranged orderly and fixed reliably. No leakage of oil, air or electricity.

5.1.7 The raw material and the purchased components of the light trolley shall comply with the relevant standards and have the certificate provided by the supplier. Important safety components shall be supplied by the qualified manufacturer and the inspections shall be carried out after their arrival.

5.1.8 Safety protection devices shall be set on the parts relating to the operation safety and all the safety protection devices shall be complete and effective.

5.1.9 The painting color of whole crane flatcar and the location of indication nameplates shall comply with the provisions of the product drawings. The coating and the inspection and acceptance of the crane flatcar shall comply with the provisions of the TB/T 2879.5-1998.

5.1.10 Horns for operation shall be installed on the crane flatcar and indication light may be installed according to the requirements.

5.1.11 The crane flatcar shall be equipped with spotlight for night work.

5.1.12 The safety of the crane flatcar shall comply with the requirements of GB/T 6067. The consumable components needed to be disassembled during operation or maintenance shall be replaceable and should be standardized and interchangeable components.

5.1.13 The protuberant parts of the track crane such as lifting hook pulley side sheathing, crane arm head, turntable rear and movable outrigger etc. shall be labeled with safety marks. The mark is the stripe in yellow and black with the angle of 45° in general and complies with the provisions of GB 15052.

5.1.14 The static deflection of the suspension spring shall comply with the design requirements after the crane flatcar assembly.

5.1.15 The radius of turntable rear shall be smaller than 2 000 m (excluding type with postpositive crane arm).

5.1.16 The grounding performance of the crane flatcar (including the crane) shall be in good condition.

5.1.17 Hand brake shall be adopted during the parking brake.

5.2 Clearance Limit Requirements

The overall dimension of the crane flatcar in servicing condition shall comply with the provisions of GB 146.1.

5.3 Running Performance of the Crane Flatcar

5.3.1 The dynamic performance of the crane flatcar shall comply with the requirements of GB/T 17426.

5.3.2 All the components shall be in normal condition and brake applying and releasing shall be normal when the crane flatcar is connected with other standard rolling stocks.

5.4 Operation Performance of the Crane Flatcar

5.4.1 The lifting capacity of the crane flatcar responding to working scope shall meet the design requirements.

5.4.2 The indicated values of all the instruments for operation control and monitoring shall be normal and meet the design requirements.

5.4.3 The brake performance shall be normal during operation and meet the design requirements.

5.4.4 The hydraulic system of crane flatcar shall comply with the requirements of GB/T 3766.

5.4.5 The lifting performance of the crane flatcar with or without outrigger on straight and flat lines or curves shall comply with the design requirements.

5.4.6 The crane flatcar shall have the function of position restriction in both left and right sections.

5.5 Requirements of Main Parts

5.5.1 Lower System

5.5.1.1 Frame (underframe) shall comply with the following requirements:

a) Reinforced beam and rotating support of the frame shall comply with the design requirements;

b) Other components of the frame shall comply with the requirements of TB/T 2187 or TB/T 2033.

5.5.1.2 Running system shall comply with the requirements of TB/T 2187 or TB/T 2033.

5.5.1.3 The components of the brake device shall comply with the requirements of TB/T 2033 or TB/T 2187.

5.5.2 Crane Operation Power System

Crane operation power system shall comply with the following requirements:

a) The main technical parameters of the diesel engine shall comply with the design requirements;

b) The installation of the engine shall have good damping function and the installation of all the parts of the diesel engine shall be firm and reliable;

c) The installation of the fuel tank shall be firm and without oil leakage or penetration. Ventilation is smooth and fuel level indicator shall be clear;

d) The capacity of the charging generator shall meet the requirements of the battery charging and DC power supply system during operation;

e) The charging generator shall charge the battery normally and charging protection device shall be installed during diesel engine operation;

f) The capacity of the battery should meet the requirements of normal starting of the diesel engine and the normal operation of the communication equipment when the diesel engine is stopped;

g) Anti-corrosion, pollutant drainage and ventilation measures shall be taken at the installing site of the battery;

h) The noise limit of the diesel engine shall comply with the provisions of GB/T 14097.

5.5.3 Electrical System

5.5.3.1 Overall Requirements

The overall requirements of electrical system are as follows:

a) The system can ensure the power supply and power capacity of all the electrical devices.

b) The system shall be able to control and protect the start-up, speed control and emergency stop of the diesel engine.

c) The system shall be able to control, interlock, protect and alarm to the hydraulic system and operation devices.

d) The system shall be able to control the associated electrical devices.

e) The system shall be able to resist the electromagnetic interference and the display of the instruments and the function of the system shall not be influenced when the onboard radio or the handset radio works.

f) The external surface of all components shall be painted evenly and consistent and the mark shall be clear and not be easily erased.

5.5.3.2 General Requirements

The general requirements of electrical system are as follows:

a) Motor, generator and other electrical devices shall have good ventilation and heat dissipation condition.

b) The electrical cables and wires shall be neatly arranged. The crimping of terminals shall be firm. There shall be clear serial numbers near the terminals. There shall be no looseness or disconnection of cables. There shall be no damage to the insulation layer of the wires and cables.

c) The installation and cabling of the console and junction box shall comply with the requirements of dust-proof, water-proof and heat dissipation.

d) The cables on the frame shall be laid in the cable conduit or groove and there shall be no oil contamination or damage.

e) The power distribution box and plates of devices such as operation console shall be installed firmly and the labels of the electrical components on the panel shall be clear.

f) The instruments, switches, buttons, indication lights, spotlights shall be installed firmly, work normally and indicate correctly. The electrical components installed outdoors shall have water-proof covers.

g) The grades and specifications of all the instruments shall comply with the design requirements and all the instruments shall be in the calibration cycle. The instruments shall be at'0'position before being powered and the measured values shall be within 20% to 95% of the range of instrument.

h) In general, the insulation resistance of the cables and wires of all the circuits to the ground shall be higher than 1 MΩ. The insulation resistance shall be higher than 0.5 MΩ when the absolute ambient humidity is higher than 16 g/m^3.

5.5.4 Illumination Device

5.5.4.1 The alarm light, operation spotlights, indication light shall be in good condition and work normally and be installed firmly.

5.5.4.2 The illumination value of the console desk in the operation room shall not be lower than 50 lx and the illumination value at the lifting hook shall not be lower than 10 lx.

5.5.5 Hydraulic System

5.5.5.1 Overall Requirements

The overall requirements of the hydraulic system are as follows:

a) The hydraulic system shall work stably without abnormal vibration or noise;

b) The hydraulic system shall have the following operation control functions:

1) Hook can lift up and down solely;

2) Crane arm can stretch out and draw back solely;

3) Crane arm can change outreach solely;

4) Upper part of crane can rotate left and right solely;

5) At least two mechanisms among the hoisting winch, extending mechanism, luffing

mechanism and slewing mechanism shall be able to operate simultaneously.

c) The hydraulic system shall be equipped with emergency pump which can draw back the operation devices to home position in 15 minutes in case of emergency and make the crane flatcar back to state of section running.

5.5.5.2 General Requirements

The general requirements of the hydraulic system are as follows:

a) The hydraulic system shall be equipped with all necessary oil filters, accumulators, circuit protection devices and so on;

b) The pressure values of all the hydraulic loops shall comply with the design requirements;

c) The operation speed, maximum speed, maximum displacement, operation pressure and maximum pressure of all the hydraulic pumps of the hydraulic system shall comply with the design requirements;

d) The displacement, rated pressure and rated speed of all the hydraulic motors of the hydraulic system shall comply with the design requirements;

e) The technical parameters, operational stroke and push and pull force of all the oil tanks of the hydraulic system shall comply with the design requirements;

f) The hydraulic system shall be equipped with safety devices to prevent overload and collision. The setting pressure of safety relief valve shall not be greater than 110% of the rated pressure and meanwhile not be greater than the rated pressure of the hydraulic pump;

g) The cleanliness, type and specification of the hydraulic oil of the hydraulic system shall comply with the design requirements and the cleanliness shall comply with the provisions of GB/T 14039-1993;

h) The pressure of all the loops of the hydraulic system shall be monitored and displayed. All the instruments shall be installed in the area convenient for the driver to observe and there shall be obvious indication labels. The accuracy of all the instruments shall comply with the design requirements.

5.5.5.3 Assembly Requirements

The assembly requirements of the hydraulic system are as follows:

a) The hydraulic system assembly shall comply with the provisions of GB/T 3766;

b) The important components of the hydraulic system shall be subject to the inspection after their arrival and retest shall be conducted to confirm its performance. They can only be installed on the crane flatcar after fulfilling the requirements;

c) Safety relief valve can only be installed on the crane flatcar after finishing the adjustment on the test bench;

d) The strength level and fastening torque of all the joint bolts shall comply with the requirements and anti-loosening components shall be ready and complete;

e) The pipe shall be located orderly and the elbow shall not be over crimped. The pipe clamp shall be arranged reasonably and fixed reliably;

f) The hose shall be located orderly and the hose clamp shall be arranged reasonably. The connecting hose with relative moving parts shall have proper length and be bound firmly without friction between them;

g) All the connectors of the pipe and hose shall be arranged reasonably and convenient for maintenance;

h) All the manual operation valves shall work reliably and flexibly.

5.5.6 Crane Operation Mechanism

Crane arm, lifting mechanism, rotating mechanism, extending and luffing mechanism, outrigger and operation mechanism shall comply with the requirements of TB/T 2187.

6 Safety Device and Accident Prevention Device

6.1 The overload protection device of the crane flatcar shall comply with the requirements of GB/T 12602.

6.2 Safety protection device of the rotatable components shall be installed correctly and firmly.

6.3 The hinged components such as brake beam and pulling rods shall be equipped with safeguard devices which can ensure the running safety of the crane flatcar in case the hinged components are damaged or fallen off.

6.4 The crane flatcar shall be equipped with load-lifting height limiting device which can reliably alarm or stop the hoisting.

6.5 The front of the turntable shall be equipped with working liaison horn. Its sound shall be different from the alarm signal and the button shall be at the position which is convenient for the operation by the driver. Transceiver may also be equipped for operation communication.

6.6 The crane flatcar shall be equipped with outreach indicator (or elevation indicator) whose reading is clear and the outreach value can be observed at any time during the operation. If the outreach is not more than 5 m, the deviation shall be no more than 100 mm; If the outreach is more than 5 m, the deviation shall not be more than 2% of the outreach.

6.7 The crane flatcar shall be equipped with lifting capacity indicator whose accuracy is not less than 5%. 16 t crane flatcar shall be equipped with lifting torque limiting device which is reliable and convenient for inspection and calibration.

6.8 The general deviation of the torque limiting device shall be no more than 8%. The defined overload alarming value shall ensure that in any case, the actual load torque shall be no more than 108% of the rated load torque at the corresponding operating condition. The crane flatcar shall be able to hoist the rated capacity slowly and reliably.

6.9 The pulley shall be equipped with a device to prevent the steel wire out of the groove.

6.10 The crane flatcar to change the outreach with steel wire shall be equipped with outreach limiting device and anti-retroverting device for crane arm.

6.11 Regarding the crane flatcar with postpositive crane arm, alarm device shall be set to the ones whose arms reach out of the radius of the turntable rear by 2 m when the arms are in the status of non-extending.

6.12 The vertical outrigger shall be equipped with mechanical lock-up device.

6.13 The crane flatcar shall be equipped with fire-fighting device which shall be installed firmly and reliably. The type and number of fire extinguisher shall comply with the provisions of the fire safety.

6.14 16 t crane flatcar shall be equipped with level gauge.

6.15 The crane flatcar shall be equipped with flatcar re-railing rigging and hook to prevent rolling over.

7 Inspection and Test Methods

7.1 Requirements of General Inspection

Visual inspection and manual inspection are adopted and the inspection results shall comply with the relating provisions of 5.1 of this standard.

7.2 Clearance Limit Inspection

The Clearance inspection shall be conducted following the provisions of GB/T 16904.1.

7.3 Crane Flatcar Running Performance Inspection

The crane flatcar running performance inspection shall be conducted following the related provisions of TB/T 2187 and TB/T 2033.

7.4 Crane Flatcar Operation Performance Inspection

The crane flatcar operation performance inspection shall be conducted following the related provisions of GB/T 6068.2-2005.

7.5 Crane Flatcar System Performance Inspection and Test

7.5.1 The inspection and test of the lower system shall be conducted following the related provisions of TB/T 2187 or TB/T 2033 respectively.

7.5.2 The inspection and test of crane operation power system, electrical system, hydraulic system, safety device and accident prevention device shall be conducted following the related provisions of TB/T 2187.

7.6 Crane Operation Device Inspection

The inspection of crane operation device shall be conducted following the provisions of GB/T 6068.2-2005.

7.7 Crane Flatcar Running Performance Test

7.7.1 Dynamic Performance Test

The dynamic performance test shall be conducted following the related provisions of GB/T 17426.

7.7.2 Trial Running Test

The trial running test shall be conducted following the related provisions of TB/T 2187.

7.8 Crane Flatcar Operation Performance Test

The idle test, crane performance parameter measurement, rated load test, dynamic load test, static load test, stability test and structure test shall be conducted following the related provisions of GB/T 6068.2-2005.

8 Test Rules

8.1 Tests include type test and routine test.

8.1.1 Type Test

Type test is to comprehensively assess whether the basic parameters, structure, performance of the crane flatcar meet the requirements of design. Test items shall be conducted on the same crane flatcar in principle.

8.1.2 Routine Test

Routine test is to check whether the structure and performance of each delivered crane flatcar comply with the type test result.

8.2 Type test shall be conducted for the crane flatcar under the following conditions. Test items shall be done according to the items marked with 'T' in Table 2, 8.4 of this standard. The test can be done by outsourcing or coordinating with the user if the manufacturer is unable to do the test by itself:

a) Newly-designed and manufactured crane flatcar;

b) Re-manufacturing after manufacturer transfers or stopping manufacturing for one year;

c) When the changes of structure, manufacturing process or material influence the performances of product.

8.3 Routine test shall be done for each massively-produced crane flatcar and test items shall be done according to the items marked with 'S' in Table 2, 8.4 of this standard. Routine test result shall comply with the type test.

8.4 The inspections and test items of crane flatcar are in accordance with Table 2.

Table2

Serial number	Inspection and test contents	Test items	Type test (T)	Routinetest (S)
1	Requirements of general inspection	5.1	T	S
2	Clearance limit inspection	5.2	T	—
3	Crane flatcar running performance inspection	5.3	T	S
4	Crane flatcar operation performance inspection	5.4	T	S
	Main parts inspection and test			
	Lower flatcar system	5.5.1		
5	Frame (Underframe)	5.5.1.1	T	S
6	Running system	5.5.1.2	T	S
7	Brake device	5.5.1.3	T	S
8	Crane operation power system	5.5.2	T	S
9	Electrical system	5.5.3	T	S
10	Illumination device	5.5.4	T	S
11	Hydraulic system	5.5.5	T	S
12	Safety device and accident prevention device	6	T	S
13	Crane operation device inspection	7.6	T	S
	Crane flatcar running performance test			
14	Dynamic performance test	7.7.1	T	—
15	Trial running test	7.7.2	T	S
	Crane flatcar operation performance test			
16	Idle test	7.8	T	S
17	Crane performance test	7.8	T	S
18	Rated load test	7.8	T	S
19	Dynamic load test	7.8	T	S
20	Static load test	7.8	T	S
21	Stability test	7.8	T	—
22	Structure test	7.8	T	—

8.5 The quality inspection department of the manufacturer shall be responsible for the acceptance of the crane flatcar before delivery and issuance of the certificate. The following technical documents and spare parts shall be delivered with the crane flatcar:

a) Product certificate;

b) Operation and maintenance manual;

c) Consumable components list;

d) Spare parts, accessories, tools and the list.

9 Marking, Packaging, Transportation, Storage

9.1 Marking

The crane flatcar shall be labeled with nameplate at its obvious position. Its dimension shall comply with the provisions of GB/T 13306. The contents shall include:

a) Product name and type;

b) Main technical parameters;

c) Serial number, date and name of manufacturer.

9.2 Packaging

9.2.1 The concomitant spare parts, accessories and tools shall be packed in box. And the box shall be firm and reliable. The label on the box shall be obvious and clear.

9.2.2 The concomitant technical documents shall be packed with moisture-proof materials.

9.3 Transportation

The crane flatcar should be transported by traction of railway motor trolley or other methods.

9.4 Storage

The crane flatcar shall be stored at the site where air is dry and circulating and free from metal corrosive gas and insulation destructive gas. The stored crane flatcar shall be maintained following the instruction of maintenance manual.